GW00739053

# Tree Dogs,

## Banshee Fingers and Other Irish Words *for* Nature

## Manchán Magan

*Illustrated by Steve Doogan*

GILL BOOKS

# About the author and illustrator

**Manchán Magan** is a writer and documentary-maker. He has written books in Irish and English about his travels in Africa, India and South America, and two novels. His most recent book, *Thirty-Two Words For Field*, explores the insights the Irish language offers into the landscape, psyche and heritage of Ireland, and he is currently working on a follow-up book. He writes occasionally for *The Irish Times*, and presents *The Almanac of Ireland* podcast about the heritage and culture of Ireland for RTÉ Radio 1. He has presented dozens of documentaries on issues of world culture for TG4, RTÉ and the Travel Channel. Having been brought up in Dublin, with long periods spent in the West Kerry Gaeltacht of Corca Dhuibhne, Manchán now lives in Co. Westmeath, in a grass-roofed house near Lough Lene, surrounded by his oak trees, and with bees and hens for company. www.manchan.com

**Steve Doogan** is an award-winning illustrator and printmaker from Scotland. He makes illustrations for all kinds of things and lives in Dublin with his wife and two daughters. www.doogan.ie

# Contents

# Introduction

WE IRISH HAVE been on this rocky green island for thousands of years, and we've been speaking our native language for a significant amount of that time. We might even have kept alive some of the words of the Neolithic people who came before us and who built the great passage tombs and portal tombs at sites like Newgrange, the Burren and the Hill of Tara.

Over the course of such a timespan we've developed precise ways of describing our surroundings, our psychology and our hopes and fears and also of describing the incredible complexity of the ecological biosphere that has sustained us almost exclusively for the majority of these millennia.

This book is about all that. It's a celebration of the wonderful linguistic legacy that we've inherited. It explores the ways in which our language deciphers and describes our world, our weather, our landscape and our reality – real and imagined.

Ultimately, its aim is to alter your perception just a little. Neuroscience tells us that a language can't change our reality, but it has also shown us that different languages allow us to see things in different ways and to focus on different things. So, we now have proof of what we always knew: every language offers a unique window on the world. The combined wisdom and life experiences of the people who developed it over thousands of years are encoded within it.

I learnt the name *sciathán leathair* ('leather wing') before I ever heard its less evocative English equivalent, bat, and I'm convinced that it had an impact on me. Likewise, my fondness for the corn bunting was influenced by hearing its Irish name first, *gealóg bhuachair* ('little bright one of the cowpat').

When you learn about the five words Irish has to delineate the stages of dawn, it changes your experience of sunrises for evermore. Or when you hear that a squid is a *máthair shúigh* ('sucking mother'), these marine cephalopod molluscs, which deposit up to seventy thousand eggs in rock holes and crevices, appear differently to you.

Colour is particularly dynamic and malleable in Irish, and when you master how it describes the tints and tones of the light spectrum, it will alter your reality significantly. Red can be either *rua* or *dearg*, and you need to be able to adjust your optical sensitivity to differentiate between the two. Green can be *glas* or *uaine* and, to make things even more rich and complex, *glas* can also be certain shades of grey that are different from the typical greys of *liath*.

Often, the Irish words for things can be more child-friendly than the English equivalents. They seem to describe an animal, thing or phenomenon with the open curiosity of a young person rather than with the categorising instincts of a neologist.

*Tree Dogs* isn't a dictionary, nor is it a nature book. Instead, it's a joyous revelry – a frolic through the wonders of the language that developed from a linguistic seed, and that meandered from the region between the Black Sea and the Caspian Sea across the continent to us over thousands of years. We moulded and honed it to capture what was most vital to us. Then, over the past two centuries, we struggled to keep it alive when powerful forces tried to stamp it out.

The words gathered in these pages, matched with evocative illustrations by Steve Doogan, should leave you with no uncertainty about the poetry, wisdom, divilment and insight contained within our glorious old tongue. Gaeilge is our birthright – something that we should be immensely proud of, not only for its cultural wealth and social and psychological subtlety but also for the insights it offers into the flora and fauna, the climate patterns, the moon cycles, the ocean currents and the otherworldly dimensions of this, our island home. Whether we pass it on as a precious heirloom or let it dissipate and die is up to us. *Is í ár dteanga í, agus beatha teanga í a labhairt.*

# WILD ANIMALS

For thousands of years, we depended on the natural resources of this island for our survival, developing our own ways of describing the wild and domesticated animals we shared it with, as well as the wild and cultivated plants that covered the land and the seabed. When we stop using these specific, local words, the countryside loses definition and specificity. Each is an individual relic of our great storehouse of accumulated wisdom. The following pages offer a small flavour of the richness of the Irish words for the animals, plants and weather phenomena on the island.

# Madra crainn
## *Squirrel*

Translates literally as 'tree dog'. The word IORA is more common for a squirrel. IORA RUA is a red squirrel; IORA GLAS is a grey squirrel. GLAS often means 'green' but not always: it can also mean 'grey' when referring to a squirrel, horse or other animal.

# Dair
## *Oak tree*

DARÓG is a small oak and RAIL is a really large one. A place that is full of oaks is DAIREACH. In Scotland, when a person was to be burnt for an evil deed, they used GLASDAIR, green oak. DAIR can also mean 'chief', because oak is the tree with the highest status in the woods. The word is found in many placenames because an oak grove was an important feature of the landscape. In the English form of Irish placenames DAIR often became 'Derry', 'dur' or 'dare', as in Durrow, Adare and Kildare.

# Cat crainn
## *Pine marten*

Translates literally as 'tree cat' and is an example of the many Irish animal names that describe how the animal looks.

# Madra allta
## *Wolf*

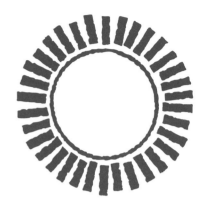

The toughest dog in Ireland was the wolf, which managed to survive on the island for a full three hundred years longer than in Britain, until the last one was killed in Co. Carlow in the eighteenth century. MADRA ALLTA translates literally as 'wild dog'. Other names include FAOLCHÚ ('wild dog'), with FAOL meaning 'wild' or 'untamed'.

The most mysterious word for the animal is MAC TÍRE, 'son of the land'. No one is quite sure why it was called this, except that wolves were believed to occasionally turn into humans, and there are many tales in Irish mythology of wolves being able to talk or act like people. Wolves were thought to have supernatural abilities, not only in Ireland but all over the world.

# Lus na mban sí
## *Foxglove*

A common garden plant with long oblong leaves and striking clusters of pendulous, bell-shaped flowers that grow along one side of a tall flowering stem. The Irish name means 'plant of the fairy women' or 'plant of the banshees', who were dangerous otherworldly beings connected to death. The plant contains a powerful poison, digitalis, which explains its association with death. The plant is also known as LUS MÓR ('big plant'), MÉIRÍNÍ NA MBAN SÍ ('banshee's fingers') MÉARACÁN AN DIABHAIL ('the devil's thimble'), MÉARACÁN DAOINE MARBH ('dead people's thimble'), MÉIRÍNÍ MADRA RUA ('fox's little fingers') and FÉIRÍN SÍ ('fairies' present'). This final name could be because the herb was believed to rescue children who had been put under a spell by fairies. Drinking the juice of twelve leaves daily revived an ailing child, principally because the digitalis chemical would strengthen the force of their heartbeat.

# Madra rua
## *Fox*

Translates literally as 'red dog'. The word SIONNACH is also used. In English 'red' is considered one colour, but in Irish we differentiate between DEARG and RUA. DEARG is the word for a dark or vibrant red, as in red ink, blood, gore, fire, embers, hot iron or the lower layers of soil. Things that were a bit more brownish-red are referred to as being RUA. So, a fox is a MADRA RUA, not a MADRA DEARG.

# Oisín
## *Fawn*

From an old word for deer, OS. A young seal is known as OISÍN RÓIN. Fionn Mac Cumhaill's son in the Fianna was named Oisín because his mother was turned into a deer by a druid, FEAR DORCHA ('Dark Man'). Oisín was reared as a deer in the wild and met his father only when Fionn's deerhounds came upon him playing naked at the age of seven with deer on Benbulbin.

## Rannán
## *Lowing of deer*

This is connected to RANNAIREACHT ('composing or reciting poetry'). LANGÁN also means the lowing of deer.

# Fia rua
## *Red deer*

A stag is a CARRIA, while a female red deer is a FEARB. A herd of deer is a DAMHRA. Other words for deer include OS, AGH ALLA ('wild cow') and FIAMHÍOL, from MÍOL ('creature') and FIA ('wild'). The word for when deer are in heat is RATAMAS, which also means being keen to work: TÁ RATAMAS OIBRE AIR.

# Fraoch
## *Heather*

This mountain plant grows only on the poorest of land and was regularly burnt back each year in FALSCAITHE, which are mountain fires, usually heather fires. A stalk of burnt heather is a SPEATHÁN, and SPEATHÁNACH is the charred stalks that were brought home to burn. FRAOCHDHAITE means 'heather-coloured' and usually refers to a colour of tweed. SOP PÍCE is a wisp of heather or of some other brush-like plant used to scour the insides of dairy vessels. There's a seanfhocal about heather that suits these days of environmental awareness: SPÁRÁIL NA CIRCE FRAOIGH AR AN BHFRAOCH IS AN SAOL GO LÉIR AG ITHE ('As sparing as the grouse with the heather while the whole world eats').

# Múirling
## *Mist*

This word is also used for a sudden heavy downpour.

9

# Slánlus mór
## *Ribwort plantain*

A common plant of meadows and wasteland with deep-veined dark-green leaves and small flowers grouped in a spike at the end of a long stem. Its name translates as 'herb of health', as it has powerful healing attributes. It will effectively heal nettle stings and insect bites (unlike applying dock leaves, which is just a placebo). The leaves can also be used to make a tea that is an effective cough medicine, and they are powerful healers of wounds. In fact, SLÁNLUS was believed to be capable of reviving the dead and is said to be the plant used to tend to Jesus's wounds on the cross. It's also called LUS NA SAIGHDIÚIRÍ ('plant of the soldiers'), because children use the flower heads in a game called soldiers in which they fence with them, trying to behead their opponents' stalks.

# Gráinneog
## *Hedgehog*

The name literally means 'ugly little one', which is unfortunate, as nothing is sweeter than a hedgehog. They also provide a valuable service by eating slugs and other pests. GRÁINNEOGACH means 'like a hedgehog', 'bristly' or 'short-tempered'.

# Comhla bhreac
## *Magic doorway*

This doorway (literally 'speckled door')
into the fairy world is found in fairy
dwellings among rocks. It's invisible
unless you know where and how to look.

# Sciathán leathair
## *Bat*

Translates as 'leather wing'. IALTÓG is
another word for a bat, and a generation
ago BÁS DORCHA ('dark death') was
used.

# Saileog
## *Willow tree*

Also called SAILLEACH or SAIL,
though this should not be confused
with another meaning of SAIL, which
is a timber beam or a cudgel, giving us
the word 'shillelagh' (from SAIL ÉILLE,
a heavy stick with a leash or thong tied
to it). The phrase BORRADH NA SAILÍ
refers to really fast growth, as willow
is the fastest-growing native tree. It's
also the most flexible and supplest
type of wood, which might be why
it was believed that placing a willow
branch over a door could make the
people inside dance without stopping.

SAILEOG is one of the SEACHT
N-ÓG NA COILLE ('seven ógs of
the wood'), alongside CUFRÓG
(cypress tree), DRISEOG
(bramble), FEARNÓG (alder),
FUINSEOG (ash), POIBLEOG
(poplar) and RAIDEOG
(bog-myrtle).

13

# Lus na gcnámh briste
## *Comfrey*

A waterside and woodland plant with big, broad, hairy leaves that produce purple bell-shaped flowers. The name translates as 'the plant of broken bones', because applying it to the body boosts the regeneration of cells in connective tissue, bone and cartilage and can even reconstruct torn and injured muscles. It's also known as MEACAN DUBH ('black root') and is widely known for its ability to quickly ease swelling.

# Iomas gréine
## *Sun inspiration*

This magical term, from Old Irish, refers to blisters caused by the sun on the leaves of certain herbs that, when eaten, were believed to give the gift of poetry.

# Asarlaíocht
## *Magic*

It was believed that magic herbs could allow you to tell the future. There are dozens of words in Irish for different kinds of magic: DATHÚ is a magical power that enables you to be dealt the best cards from the pack; TUAITHLEASÓG refers to a girl who can cast magic spells; COSTA AGUS SCÁRAOID are a magic goblet and tablecloth that produce all the drink and food you could ever desire; and BIORÁN SUAIN is a magic pin that summons sleep.

# Easóg
## *Stoat*

The word derives from EASCANN, an eel, because of the stoat's wriggling body and sinewy shape. IS Í AN EASÓG Í ('She is the stoat') is used to refer to someone hotheaded or spiteful. BEAINÍN UASAL ('small lady') is another word for the animal, as is CAILÍN BÁN ('darling girl') and FLANNÓG.

15

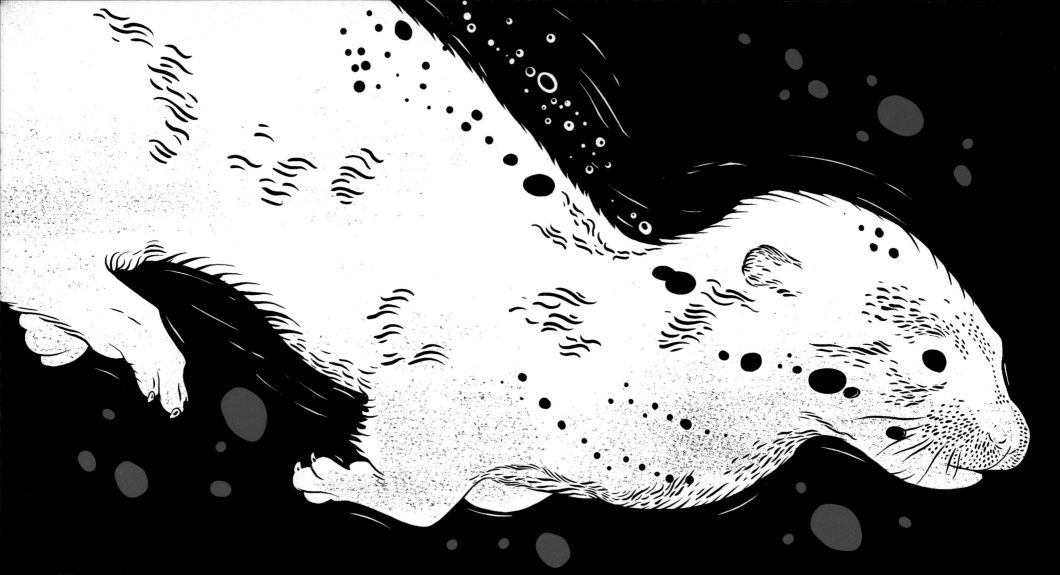

# Dobharchú
## *Otter*

Translates as 'water hound', as opposed
to DOBHAR-EACH, which translates as
'water horse' and means hippopotamus.
An otter is also referred to as a MADRA
UISCE ('water dog') and as a DOBHRÁN
or CÚ DOBHRÁIN.

# Earc luachra
## *Lizard*

Translates as 'lizard of the reeds', as
opposed to EARC SLÉIBHE, which is a
newt and means 'lizard of the hills'.

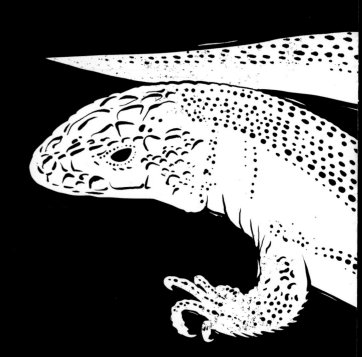

# Riathróir

## Sure-footed animal on boggy ground

Other words for being nimble or sure-footed are LUAINEACH ('nimble like a horse'), SCODALACH ('good at scuttling') and OSCARTHA ('lithe' or 'agile'). A light, nimble person can be called a CLEITIRE ('feather-like person'). The opposite is CAM REILIGE ('bandy', 'club-footed'), which literally means 'crooked from the graveyard' and comes from a superstition that a pregnant woman treading on a grave will give birth to a club-footed child.

# Seilín cuaiche

## Meadow cress, large bitter cress

A flowering plant of damp meadows and bog gardens. It translates as 'little spittle of a cuckoo', perhaps because it appears in late spring and early summer, when the cuckoo begins to call. It shouldn't be confused with SEILE CUAICHE ('cuckoo spit'), the blobs of white froth made by a sap-sucking insect in springtime.

# Stuaicín
## Cow with upturned horns

There are many words to describe different types of cow, such as CÚBACH, a cow with bent-in horns; DROIMEANN, a white-backed cow; RIABHACH, a speckled cow; BRAOBAIRE, a reckless cow (or a rude person); CEANNANN, a white-faced cow (or a star-shaped blaze on an animal's forehead); CEART-AOS, a two-year-old heifer; COLANN, a yearling cow; CNÁMHARLACH MAIRTE, a bony, withered cow; BLÉINEACH, a white-loined cow; BRADAÍ, a thieving or wandering cow (or person); BOINÍN, a little cow; BLEACHTACH, a milch cow; LOILÍOCH, a cow after calving; MAOILÍN, a hornless cow; RASPA, a bony old cow; BÓ THÓRMAIGH, a cow about to give birth; SPANGAIRE, a barren cow; BEARACH, a young heifer; LIATH, a grey cow; and BODÓG, a young heifer.

# Bainne bó bleachtáin
## *Cowslip*

A primrose-like plant of meadows and woodlands with bright-yellow flowers. Its name translates as 'milk (or juice) of the milking cow'. Cowslip flowers were rubbed on a cow's udder on May Day to protect the milk from harm by the fairies. The flowers are also used to make a herbal juice or tonic and a tasty wine.

# Athair thalún
## *Yarrow*

A liquorice-smelling grassland plant with clusters of white flat-topped flower heads. Its name translates as 'father of the earth', because it was considered one of the most powerful plants for healing, soothsaying and protection from evil forces or maybe because yarrow grows in gravelly ground and, after a decade or more, soil builds up around it, allowing grasses to colonise the newly created earth.

An honoured elder among the herbs, it can lower blood pressure, revive digestion, heal bruises and staunch bleeding wounds. It can also restore the 'goodness' to milk that has had its butter-making abilities robbed from it by fairies. The plant was also known as LUIBH NA NDAITHEACHA ('plant of rheumatism'), for its ability to dramatically ease the pain of rheumatism and arthritis.

# Damh
## *Ox*

An ox is a bull that cannot father calves. Placenames containing the word DAMH are much more common than those containing the word TARBH, which is a bull. A DAMH was a valuable animal, as it could plough and pull carts. The Old Irish word DAM CONCHAID ('wolf-fighting ox') referred to an animal brought along with a herd of cows to the summer pastures to defend against attack by wolves.

# Drúchtín
## *Sundew*

The name of this plant also means 'good fortune' and 'finding the love of your life', because it was believed that a girl would discover the hair colour of her future husband from the shade of the colouring of the morning dew on May Day.

# Capall
## *Horse*

CLIBISTÍN is a shaggy horse, FALAIRE is an ambling horse and GEARRÁN is a gelding or packhorse (or a strong-boned woman). The verb for falling off a horse (or from a tree or anything in which you grip on to) is ASCAR. EACH is another word for a horse, as is PEALL, which is also a pelt or pieces of coarse cloth. MARC is yet another word for horse, as in the phrase AR MHUIN MHAIRC A CHÉILE, all higgledy-piggledy (literally, 'on the back of each other's horses'). GRAFAINN is a horse race or a group of men devoted to horses. GRAIFNE is horse racing and GRAIFNEACH is fond of horse-racing … or the grunting and squealing of animals. GRAÍ is a stud of horses, as is EACHRA. HOBAIREACHT is the act of calling to a horse, such as calling, HOB AMACH! ('To the left!'). We've already seen that LIATH means 'grey', but it can also be a grey horse or grey cow.

## Seitreach
### *Neighing, braying*

The sound horses make when they meet after an absence. It is also the cry of a hawk in hunting and the sound of grumbling and complaining.

## Staigín
### *Useless horse or person*

This word is connected to STAGA, a potato softened and spoiled by frost or otherwise rotten and worthless.

## Plúirín *Violet*

Also means 'little flower' or 'pretty girl'. SAILCHUACH also means 'violet'; it translates literally as 'cuckoo stain'.

# Gabhar
## *Goat*

POCÁN is a he-goat, from POC, which is the bucking of a goat or deer. FEARPHOC is another word for a he-goat, which can be translated as 'bucking man'. A castrated he-goat is a COLLPHOC. MINSEACH is a nanny-goat and FUAIRNEACH is a barren goat (or a cold, unemotional person). MEIG is the bleat of a goat and MEIGEALL is a goat's beard. MEIGEALLACH means being bearded like a goat, bleating like a goat or talking foolishly. The phrase CÁ BHFUIL AN GABHAR Á RÓSTADH? (Literally, 'Where is the goat being roasted?') means 'Where is the fun to be had?' (Fun as in FRAECSÁIL, which means acting the giddy goat.)

# Sliomach
## *Unripe potato*

Also means 'useless person'.

# Gionán
## *Small potato*

Also means 'insignificant person', as do CREATHÁN, BODALÁN and CLAMHRÁN. A large potato, or a darling person, is a GILLÍN.

# Gabhairín
## *Small goat*

A word also used for potatoes that were sold secretly by children for pocket money.

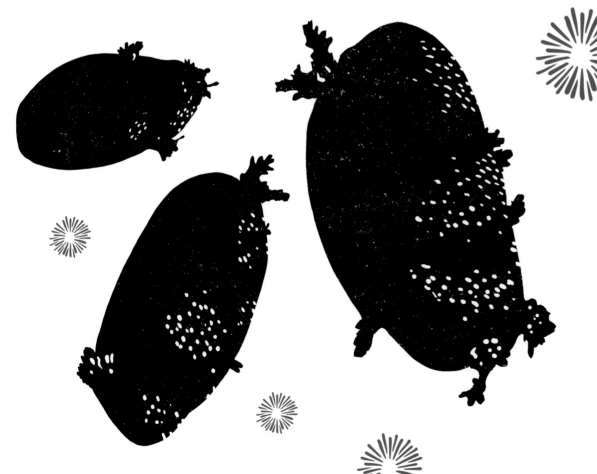

# Falaireacht
## *Ambling gait of a spancelled goat*

A spancel is a rope that ties together the legs of an animal to stop it wandering. There are many words for different types of spancel, such as BUICMÍN, SPEIRSÍN, BUAIRCÍN, CRANN-NASC, BUARACH and URCHALL.

# Muc
## *Pig*

A sow is a CRÁIN, and a sow in heat is a CLÍTHSEACH, which is also a disrespectful term for someone. MUCACHÁN is a pig-like person. HURAIS is what you say to call pigs. You can also say, TOCH! TOCH!

# Scimeáil
## *Prowling*

A word used for a pig or dog that is prowling for food, loafing or watching for a drink. The word is a borrowing of the English word 'skim'. SIRTHEOIREACHT is also prowling and foraging. SMÚRTHACHT is nosing and sniffing around, and BRANAIREACHT is the act of prowling for prey.

# Capánach
## *Little pig fed with milk from a saucer*

A CRAMPÁNACH is an underdeveloped piglet, and an ARCÁN is a stunted little pig who sucks the hindermost teat of all. A COPSAÍ is an extra piglet whose mother has not got enough teats on her to feed, and an ÍOCHTAIRÍN is a spoon-fed suckling pig.

# Feimide
## *Pig's curly tail*

A word that comes from FEIMÍN, which is a little tail or a tuft and also the curled feathers in the tail of a drake (male duck). FEIMÍN derives from the word FEAM, which is the rubber-like stump on seaweed. FEAM is also the origin of the word FEIMINEACH, an animal that bites the tails of other animals. The word is mostly used to refer to cows, though pigs can also bite each other's tails when bored.

# IN THE AIR

The Irish sky is not particularly different from that above other countries, maybe just a little cloudier, but our way of seeing the clouds and the birds and the wind they exist alongside is unique. In the following pages we will explore these elements and the insights the language offers into the sky realms. We investigate why a snipe is a 'baby goat of the air', why a blackberry means 'nothing' and why a little grebe is a 'clumsy-footed one of the wave'. We've all felt the many types of wind in Ireland. If you want to learn the words for them, read on.

# Dreoilín ceannbhuí
## *Goldcrest*

The Irish name for the goldcrest is either DREOILÍN CEANNBHUÍ ('golden-crested wren') or DREOILÍN EASPAIG ('bishop wren') – a reference to the fact that this bird's crest (the display feathers on the crown of its head) are like the ceremonial headgear of a bishop. Another name for the goldcrest is DIARMAIDÍN RIABHACH ('striped little Diarmaid'). The names of birds in Irish often contain people's names.

# Dreoilín
## *Wren*

The Old Irish name was DREOLLÁN or DRÚI DONN ('brown druid'). In Welsh DRYW means both 'druid' and 'wren'. The wren was considered a magical bird that had powers associated with the gods. It was hunted in winter because it represented the old year, while the SPIDEOG (robin) brought in the new year. Capturing and killing a wren on St Stephen's Day meant you could take on its powers. DREOILÍN TEASPAIGH means 'wren of hot weather' and is the Irish word for grasshopper.

# Sméar
## *Blackberry, berry, bramble, blur*

Used with a negative GAN, NÍ or NÍL, it means 'nothing': FIÚ DÁ MBEADH BOSCA MÓR SNICKERS AIGE NÍ THABHARFADH SÉ SMÉAR DOM ('Even if he had a box full of Snickers he'd give me nothing').

# Lasair choille
## *Goldfinch*

One of many Irish bird names that paints a vivid picture. It translates as 'woodland flame', which is a good description of this highly coloured bird with a bright-red face and yellow wing patch that tends to favour areas with bushes and trees.

# Rua-ghaoth
## *Squall or sudden gust*

This word can also mean an east wind. CUAIFEACH is also a sudden gust of wind. FLEÁ means the same thing but often with rain in the gust. SAIGHNEÁN also means a blast of wind, but it too can mean other things, such as a thunderbolt, lightning or a hurricane.

# Rí rua
## *Chaffinch*

Ireland's most common finch is the chaffinch, which in Irish is RÍ RUA ('red king'), with the RUA referring to the pinkish orange-brown of the male's breast, face and underside.

33

# Smóilíní
## *Young thrushes*

CHUIRFEADH SÉ NA SMÓILÍNÍ AG SCLIMPIREACHT I DO CHROÍ ('It would set the young thrushes dancing in your heart').

# Sclimpireacht
## *Sparkling*

Also means 'glancing' or a form of joyous dancing.

# Fearnóg
## *Alder tree*

Also known as FEARN, which can mean a sailing mast, as masts were mostly made of alder long ago. RUAIM is another name for the alder tree, and this word is also used for a dark-red dye made by boiling alder bark, sorrel, briar roots and dock roots together. FEADÁN is the pith of the alder tree and it also means a reed or a tube or pipe.

# Eireaball cait
## *Bullrush or reed mace*

Literally, 'cat's tail', it is a native pond plant with long, flat greyish-green leaves and brown sausage-like flower heads on tall, strong stems. The plant is also known as COIGEAL NA MBAN SÍ ('banshee's distaff', a distaff being a stick onto which wool or flax is wound in spinning). Bullrush wasn't much used for herbal medicine, but it was believed to be a fairy plant and so was feared. The roots, stems and young shoots are edible and nourishing, containing sugars and protein, like corn on the cob.

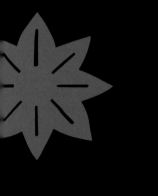

# Ceann cait
## *Long-eared owl*

Translates literally as 'cat's head', because of its round face and tufted ears. The main word for an owl is ULCHABHÁN, though they were also known as SCRÉACHÓG REILIGE ('graveyard screechers') and CAILLEACH OÍCHE ('witch of the night'), because of the way they suddenly appear flying low and near you through the darkness.

# Meannán aeir
## *Snipe*

A wading bird with a long, straight bill, it is found in marshy vegetation. The name translates as 'sky kid' or 'baby goat of the air', because of the eerie goat bleating produced by its feathers, which stick out at the tail sides. They vibrate as the bird flies in a roller-coaster pattern. A male snipe is known as GABHAR DEORACH ('crying goat'), possibly because of this sound or its repetitive *chipper, chipper* song.

The bird is also known as a NAOSCACH and a BODACHÁN, which means 'little oaf' or 'clown', possibly because of its annoying bleating. There's a phrase AN FAD BHEIDH NAOSC AR MÓIN NÓ GOB UIRTHI ('As long as there's a snipe in the bog or a bill on the snipe'), which means 'until the end of time'. But snipe numbers have declined significantly in recent decades because of the draining of moorlands and the foresting of bogs, so this may no longer be a great way to gauge the endlessness of time.

# Lus an bhainne
## *Milkwort*

A low-growing, sometimes trailing flower of grasslands that looks a bit like a baby bluebell. Its name means 'plant of the milk', a reference to the traditional belief that it could help a mother produce more milk after childbirth. It's also called LUS NA SEACHT MBUA ('herb of seven gifts'), because of its varied medicinal uses. Another name for it is NA DEIRFIÚIRÍNÍ ('the little sisters'), because it can have any of four coloured flowers: blue, purple, pink or white.

# Spágaire tonn
## *Little grebe*

A small, dumpy water bird. A SPÁGAIRE is a person with big, clumsy feet or an ungainly way of walking, and TONN means 'wave'. The little grebe does have big feet or, rather, cartoonishly enormous ones. These look comical and seem clumsy on land, but in fact they are a superb adaptation to the bird's aquatic lifestyle and enable it to do some seriously impressive diving.

# Nead
## *Nest*

Also a portion of hay or corn that is a different colour from the rest. NEAD can refer to all kinds of nests, from floating ones built by little grebes on a raft of waterweeds to a golden eagle's eyrie. The act of nest-building is BROBHAÍL or SOPAIREACHT, from BROBH and SOP, which mean 'wisp'. NEADAIREACHT is another word for nest-building. FOIRGNEAMH is the principal word for a building, but it also refers to a nesting place, such as on the water's edge or on a cliff face.

# Snag
## *Lull*

Easing of the wind, lull in a storm.
It brings on a SOIRBHEAS, a fair sailing
wind (or an easy life): THÁINIG SNAG
BEAG AGUS THUG SOIRBHEAS LEIS
('The storm eased and brought a fair
wind').

## Lacha mhásach
### *Pochard*

A type of duck that is quite rare now in Ireland. The name translates as 'duck with large buttocks'. MÁSACH can be used to refer to any big-bottomed or big-thighed animal and is generally a compliment – but less so when used for people.

## Tonóg
### *Duck, stout little woman*

LACHA more commonly refers to a duck and can mean 'fair young girl'. The collective word for ducks is LACHAR, and a place that is abounding in them is known as LACHNACH. To call ducks to their food, you would say, FÍNEACH! FÍNEACH! They also respond to HUÍT, HUÍT.

# Lus gan athair gan mháthair
## *Duckweed*

An aquatic plant consisting of thousands of tiny rounded leaves floating on the water. The name translates as 'plant without father or mother', because they reproduce by budding out of themselves rather than from seed produced by a parent plant. Another word for duckweed is ROS LACHAN ('duck seed'). It was believed in the tenth century that the word ROS came from RÓ-FHOS, meaning 'great rest', as the plant is always seen resting idle on top of stagnant or slow-moving water.

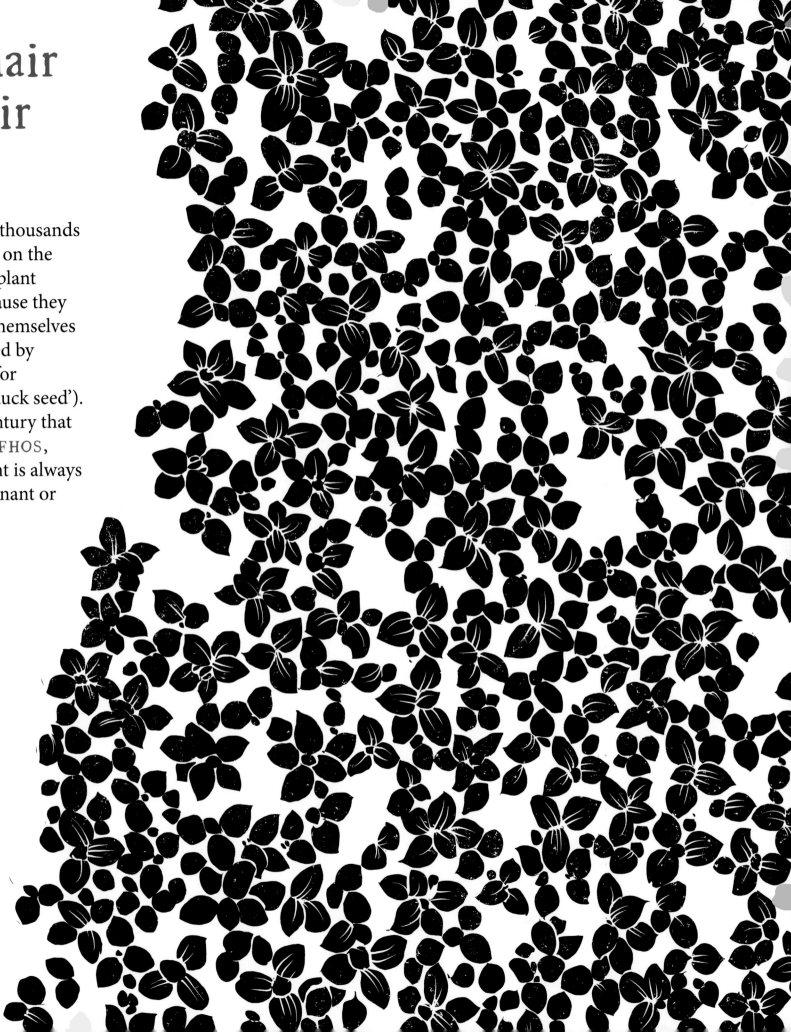

# Gealóg bhuachair
## *Corn bunting*

A sparrow-sized songbird of hedgerows and farmland. The name translates as 'little bright one of the cowpat'. This short, stout bird with dark-brown feathers edged with grey and off-white underparts is not particularly bright in colour, but it does like to hang around cattle yards and grazing fields, where it feeds on insects and seeds. Or at least it did until it went extinct in Ireland as a breeding species twenty years ago.

# Scráib
## *Rush or gust of wind or rain*

It's similar to ROIS, which is rain driven furiously by the wind.

# Ropadh
## *Blast of wind*

The word is also used for a fight, a bursting or tearing through, a sudden or violent putting away and a row. FUARGHAOTH is a cold blast of wind, IOMGHAOTHACH is an eddying blast of wind and SAIGHNEÁN is a sudden blast of wind, although in the plural, NA SAIGHNEÁIN, the word means 'the Northern Lights'.

# Seabhach
## *Hawk*

In poetry SEABHACH is often used to refer to a champion or warrior. A perceptive person is referred to as SEABHCÚIL ('hawk-like'), as opposed to SEABHCÓIR, which is a person who hunts with a hawk, or a fowler. A SEABHCÁN is where a trained hawk is kept, and the act of hunting with it is SEABHCÓIREACHT. A kestrel is SEABHAC GAOITHE ('wind hawk'). A phrase meaning to tear something into bits is GREAMANNA SEABHAIC A DHÉANAMH DE RUD (literally, 'To make hawk bites out of something').

# Crann creathach
## *Aspen*

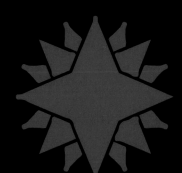

Translates as the 'quaking (or trembling) tree', because the leaves of the aspen are constantly quivering, even in the most meagre breeze. In fact, they shake when there's virtually no wind at all, making it seem as if the tree is trembling of its own accord. The leaves make a lovely, relaxing sound as they shake and shiver, and they have a sweet smell in spring.

# Pilibín míog
## *Lapwing*

MÍOG is a bird's 'cheep', so the name
literally translates as 'cheeping plover',
because plovers are just like small
lapwings, and the fact that they belong
to different groups within the same
subfamily of medium-sized wading birds
wasn't known. The PILIBÍN MÍOG has
a shrill, wailing cry, even more plaintive
than the plover's, hence the name. MÍOG
is used for any bird's cheeping, but it
refers especially to the cry of the plover
and lapwing.

# Cam an ime
## *Buttercup*

A bright-yellow wild flower that grows
in fields. CAM means 'melting pot', and
IME means 'butter'. The flower looks
buttery but the name might also refer to
the fact that buttercups, like cowslips,
were rubbed into the udders of cows
to make sure they would produce rich
cream and butter. The plant is also
known as FEARBÁN, from the word
FEARB ('weal' or 'welt'), because of the
sharp, biting sap in its stem. FEARB,
as we saw earlier, is also a word for a
female deer. A final name for this flower
is CROBH PRÉACHÁIN, which translates
as 'crow's claw', because the shape of the
leaves is a little like a bird's foot.

# Pilibín
## *Plover*

Translates as 'little Philip' or 'little thingamajig'. A way of saying someone is deliberately evading an issue, or setting a red herring, is IS É TIONLACAN AN PHILIBÍN ÓNA NEAD AIGE É ('He is accompanying the plover away from his nest'), as they are known to fly off, crying with anxiety, pretending that they have a broken wing or are incubating eggs elsewhere, thereby luring predators away.

# Séamas rua
## *Puffin*

The name means 'Red Séamas'. Another word for this bird is FUIPÍN.

# Faoileán
## *Gull*

Translates as 'one who circles or spins, or wild one'.

# Muirbheach
## *Sandy soil by the seaside*

A level stretch of sandy land along the seashore.

# Stadhan

A flock of mixed gulls over shoals of fish.

# Faoileanda
## *Gull-like*

Means 'sweeping', 'graceful', 'brightest appearance'. It's connected to FAOILEÁNACH, which means 'frequented by gulls'.

49

# UNDER
# THE WATER

As an island people we have depended to an enormous extent on the seas surrounding us. We thrived in a sustainable way by living in harmony with the waves and currents, the sea creatures and plants, the rocks and strands. Over millennia we have amassed dozens of names for types of waves, winds, seaweeds, shellfish and fishing conditions. The sheer specificity of some is bewildering, such as the word for a young seal after shedding its white baby coat, or the word for a lull in a rainstorm, or the sea phosphorescence used as markers by fishermen. Curious? Dive in.

# Deilf
## *Dolphin*

The word MUC MHARA is also used to describe a dolphin, but officially it means 'porpoise'. MUC MHARA translates literally as 'sea pig'. Porpoises used to be called TÓITHÍNÍ MUCA MARA, with TÓITHÍNÍ referring to overweight people or dolphins, so TÓITHÍNÍ MUCA MARA means 'overweight dolphins'.

# Cluiche
## *Shoal of fish, a game*

CLUICHE GAINÉAD is a flock of gannets over a shoal. BÁIRE is also a shoal of fish and a match or contest. PÁIRC refers to a field or a games park, but in the phrase PÁIRC ÉISC it means a large shoal of fish. Another word for shoal is RÁTH, as in RÁTH SCADÁN, a shoal of herring. RÁTHAIGH is a verb meaning to shoal fish, like you would herd sheep. In fact, the verb CLUICH is used to mean gathering up fish in shoals or rounding up cows or sheep. It also used for the act of dogs hunting in packs or turning a hare. A dense shoal of fish, like those that have been corralled together by whales or dolphins, is a BRÓ ÉISC, and a small shoal is a TEADHALL.

# Bruth
## *Surf*

Surf on the sea, beach wrack, heat, a rash, soft downy hair or beard or downy hair on a woman's face.

# Bainirseach
## *Female seal*

CRÁIN RÓIN also means a female seal. The rookery where they have their babies is ADHBHA RÓN. A seal's lair is DALLÓG RÓIN and a shoal of seals is CLADACH RÓNTA. There's a blessing MAITHEAS BAINNE CÍOCH AN RÓIN GO NDÉANFAIDH SÉ DUIT ('May it do you as much good as seal's milk'), for it was believed that a young seal puts on a layer of fat each time it nurses on its mother. GILLÍN RÓIN is the term for a fine, plump seal. Long ago these were singled out to be hunted with a spear known as a RÓNGÁE in Old Irish. A way of expressing the idea of a time so long ago that no one remembers it is Ó BEARRADH NA RÓNTA ('When seals were shorn').

# Smugairle róin
## *Jellyfish*

Translates literally as 'seal snot'. SMAOIS, SMUGA, SPROCHAILLE and SCAMAL are all words for snot. A snotty-nosed person is a SMUGACHÁN.

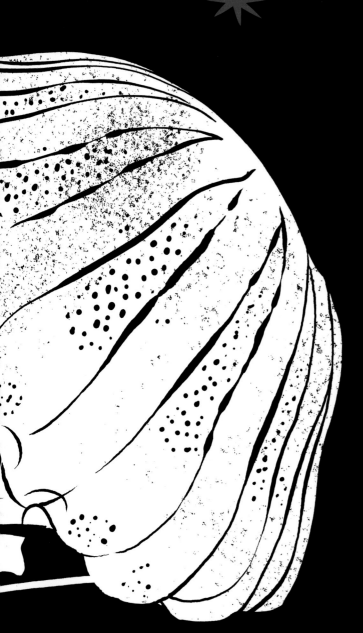

# Gormánach

## *Young seal after shedding its white baby coat*

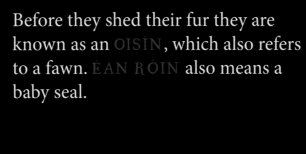

Before they shed their fur they are
known as an OISÍN, which also refers
to a fawn. ÉAN RÓIN also means a
baby seal.

# Máthair shúigh
## *Squid*

Translates literally as 'sucking mother'.
CUDAL MÉARA was also used, which
means 'fingered cuttlefish'.

# Sraith
## *Stretch of floating seaweed*

Also means 'spreading ground', 'swath
of grass' and 'quartering of soldiers'.
SRAITH ÉADAÍ is clothes spread out on
the ground to dry.

# Rúplach
## Long string of seaweed

The word also refers to roots running far into the ground or to a strong fellow, especially a tall, bony one – or to anything strong.

# Cuán mara
## *Sea urchin*

CUÁN means 'a pack of young pups, cubs or wolves' and MARA means 'of the sea'. You can also call the animal GRÁINNEOG THRÁ ('strand hedgehog') or CARBHÁN CARRAIGE, with the word CARBHÁN seeming to refer to sprat or carp rather than to a caravan, and CARRAIGE meaning 'of the rock'.

# Monghar
## *Roaring*

A noise like that of the sea. A terrible din or roar, MONGHAR NA LAOCH being the confused shouts of warriors.

# Glas
## *Green, grey*

Long ago we decided that the sea wasn't blue but was instead GLAS, by which we meant 'greeny-grey'. GLAS is used for the green of grass, leaves, young plants and other natural things, while UAINE is the word for manufactured green things. But GLAS can also mean 'grey' when referring to the colour of animals. It's also the colour of undyed wool, homespun cloth, iron, a cold winter sky and grey eyes.

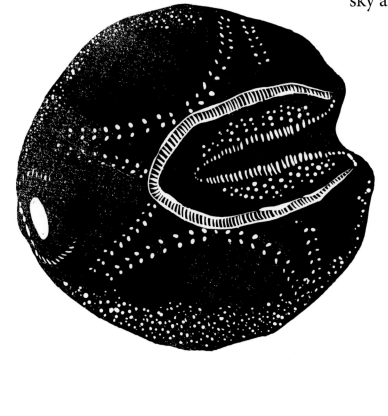

# Scadán
## *Herring*

Also refers to a thin man or to mottles in the shins from standing too close to the fire. A male herring is SCADÁN LÁIBE and a female is SCADÁN NA BPIS. SCADÁN GAINIMH is a sand-eel and SCADÁN CAOCH, which means 'blind herring', refers to the cheapest type of sauce, made with water and salt and used to dip potatoes in. If you were a bit wealthier, you'd slice an onion and add that to help flavour the water.

# Sámhnas
## *Lull in a rainstorm*

Also means 'ease' or 'respite'. BEIDH SÁMHNAS AMÁRACH means 'The bad weather will ease tomorrow.'

# Leathóg
## *Flatfish*

LEATHÓG MHUIRE, a halibut, translates as 'flatfish of the Virgin Mary'. LEATHÓG BHÁN is white sole and LEATHÓG BHALLACH is plaice. LEATHADH NA LEATHÓIGE ORT means 'May you be as stretched out as a flatfish'. The phrase is also used to describe someone dripping wet. Any famished-looking creature or flat, flaccid thing can be called a LEATHÓG. It's also one of SEACHT N-ÓG NA FARRAIGE ('seven ógs of the sea'), alongside CADÓG (haddock), CNÚDÓG (gurnet), CRÚBÓG (spider crab), DALLÓG (dogfish), FEANNÓG (whiting) and GLASÓG (coalfish).

# Bradán
## *Salmon*

EO, ÉIGNE and MAIGHRE are other words for BRADÁN, MAIGHRE also meaning a handsome, healthy person or a beautiful woman and ÉIGNE meaning a champion. CORRÁNACH is a male salmon and DIÚILÍN is a young salmon. A river abounding in salmon is MAIGHREACH, BRADÁNACH or ÉIGNEACH, although salmon are so scarce now that we rarely get to use these words. SAOTHAR is the hole made by salmon in the sandy riverbed for depositing spawn.

# Sopóg
## *Torch*

A torch made of straw, rushes or bog deal mounted on a pole and used by poachers on the river. It is also a handful of straw and a sheaf of corn.

# Bual-lile
## *Water lily*

An aquatic plant with pinkish, floating flowers and large, round green waxy leaves. The first part of the word comes from BUAL, an old word for water that you don't see much now, except in ROTH BUAILE, a water wheel. PÓICÍNÍ LOCHA is another word for BUAL-LILE, with PÓICÍNÍ meaning little pockets or tiny patches of ground, and LOCHA meaning 'lake'. You can also call it DUILLEOGA BÁITE ('sunken leaves') or BIOR-RÓSANNA. BIOR normally means 'spike' but in this case is another old word for water, and RÓSANNA means 'roses'. Water-lily seeds were once an important part of the diet of our hunter-gatherer forebears here in Ireland.

# Míol mór
## *Whale*

Literally means 'big animal'. MÍOL refers to an animal, insect or other creature. It's the root of the word MÍOLTÓG ('little animal'), which is a midge. The phrase NÁ DÉAN MÍOL MÓR DE MHÍOLTÓG means 'Don't make a fuss out of nothing' (literally, 'Don't make a whale out of a midge').

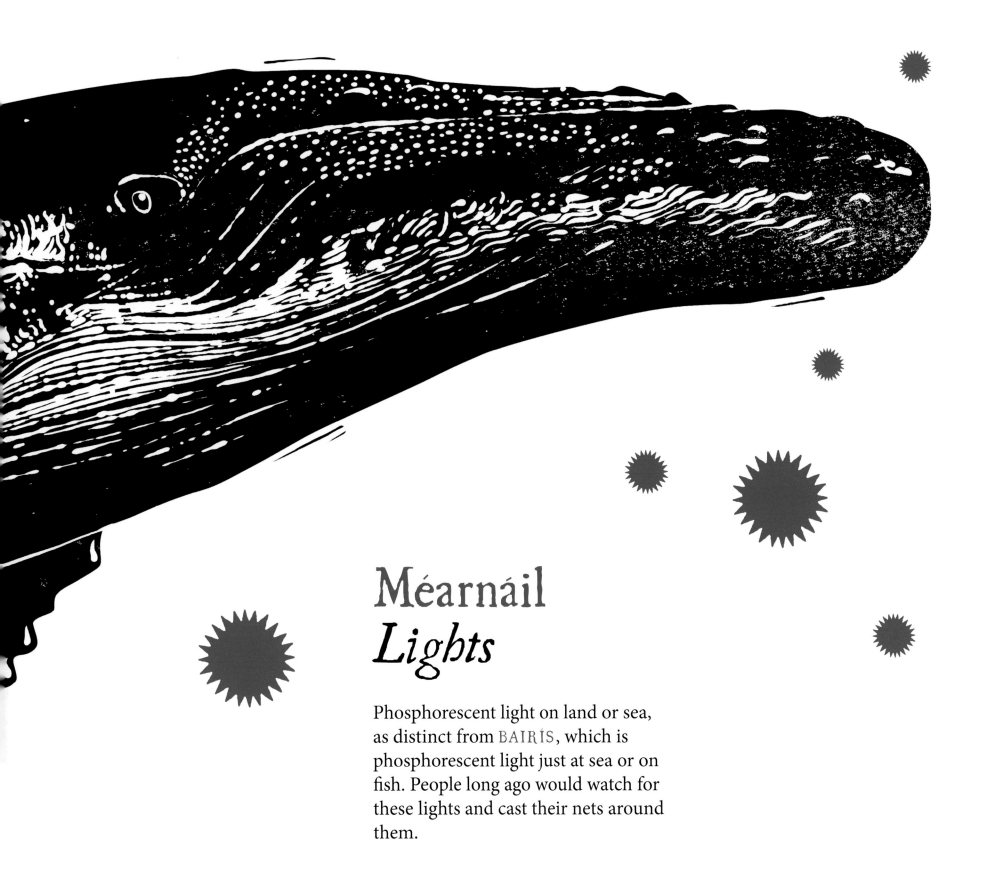

# Méarnáil
## *Lights*

Phosphorescent light on land or sea, as distinct from BAIRIS, which is phosphorescent light just at sea or on fish. People long ago would watch for these lights and cast their nets around them.

# LITTLE CREATURES

Gradually, we are coming to realise that the world depends on little things: insects, spiders and mites, as well as mosses, microbes and tiny plants that we tend to ignore as we go about our lives. Irish has kept careful note of them all, and as we rush for higher definitions on our cameras and screens, it is reassuring to see how the language can focus so sharply and deeply on the minutiae that make up the macrocosm. These words remind us of just how wide, deep and nuanced our forebears' understanding of the world was.

# Seangán
## *Ant*

Also means 'slender' or a worthless, weak person. The phrase MARCACH AR SHEANGÁN ('rider on an ant') is a way of describing someone in a ridiculous situation. CRÁINSEANGÁN is a queen ant.

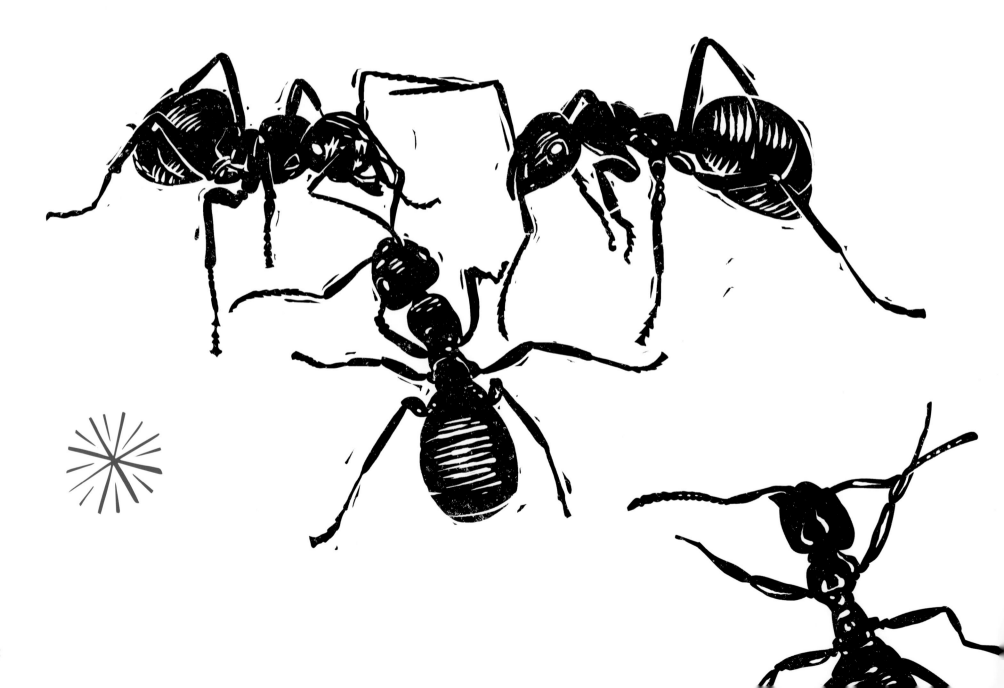

# Póiríní seangán
## *English stonecrop*

A low-creeping succulent plant found on dry walls, rocks and cliffs near the coast. The name translates as 'ants' marbles' or 'ants' jackstones', because the stubby, greyish-green leaves look like pebbles and tend to grow in stony, sunny areas that ants like. People must have imagined the ants using the roundish leaves for games. The leaves can turn pinkish in dry conditions.

# Speig neanta
## *Hairy caterpillar*

Also used to refer to an evil person. The word SPEIG appears to mean 'species', and NEANTA means 'of the nettle', so 'nettlesome being' is what it conveys. Other words for caterpillar include MÁIRÍN AN CHLÚIMH ('Máirín of the furry coat'), DÓNALL AN CHLÚIMH ('Dónall of the furry coat') and SIOBHÁINÍN AN CHLÚIMH ('little Siobhán of the furry coat'). There is also PÉIST CHAPAILL ('worm horse'), which is the black-and-yellow cinnabar caterpillar.

# Dril
## *Droplet*

A drop glancing in the sun, a twinkle.

# Deannach
## *Moss*

Moss at the bottom of wells, the hair that grows on a person's body, dust.

# Cráinbheach
## *Queen bee*

Literally 'bee sow' or 'breeding female bee'. She leaves the nest at most twice in her life: once to mate and a second time to abandon the nest with a swarm of bees, leaving a new young queen at home. If two swarms leave a nest in the same year, the second swarm is called a TARBHSHAITHE ('bull swarm').

# Cuasnóg
## *Honeybee nest*

A nest of honeybees found in a hedge or in the grass, as opposed to a TALMHÓG, which is a nest found in the ground. CUASNÓG also means 'surprise gift', as if you would find the bees and get their delicious honey and a chance to capture the nest and move it to a hive, where you would harvest honey year after year. The similar word CUASÓG refers to the honeycomb found in the nest and also to a small cavity or little hollow.

# Macshaithe
## *Bee swarm*

Literally 'son of the swarm', it is a swarm of bees derived from a parent swarm in the first year of its existence. Similarly, MACASAMHAIL means 'copy', 'reproduction' or 'likeness' and derives from MAC SAMHLA, 'son (or descendent) of the likeness'. MAC IMRISC means 'pupil of the eye' and translates literally as 'son of the iris of the eye'. MACLEABHAR ('son of the book') is a copy of a book. An old word for a young nun is MAC-CHAILLEACH ('son of the veiled woman'), and MAC AN DABA is the middle finger. MACALLA means 'echo' and derives from MAC AILLE (or MAC FHALLA), meaning 'son of the cliff (or wall)'.

# Siansán
## *Humming*

A buzzing or humming noise or a breeze. SIANSÁN NA GCON AR SLIABH means 'The crying of hounds on the moorland'. SIANSÁN is not quite the same as TORMÁN, which is a roaring sound, a rumbling noise or a whirlwind.

# Bóin Dé
## *Ladybird*

Translates as 'God's little cow', possibly because the spots on its back are similar to the patterns on some cows. The Welsh name is BUWCH GOCH GOTA, which translates as 'short red cow'; you can also use the name BUWCH FACH ADDA ('Adam's little cow'). In Russian it's BOZHYA KOROVKA, which means the same as the Irish name. The English name, ladybird, is a reference to the Virgin Mary, Our Lady. In Danish its name is MARIEHONE ('Mary's hen') and in German it's MARIENKÄFER ('Mary's beetle'). In Spanish it's MARIQUITA and also VACA DE SAN ANTÓN ('St Anthony's cow'). In Irish it's also known as BÓ SHAMHRAIDH ('summer cow'), CIARÓG NA MBEANNACHT ('beetle of the blessings') and CEARC MHUIRE ('the Virgin Mary's hen').

# Damhán alla
## *Spider*

Translates literally as 'wild little ox'. The name has been used in Ireland for at least 1,200 years. In medical texts from the fourteenth century there is advice to drink powdered DAMHÁN ALLA dissolved in water as a cure for epilepsy. The 'little ox' is also referred to in the name of the water spider, DAMHÁN UISCE.

# Luibh na seacht ngábh
## *Wall-rue*

A small tufted fern with shiny, deep-green, slightly fleshy leaves that are round, diamond-shaped or fan-shaped. It is found on walls and in rock crevices. Its Irish name is literally 'plant of the seven perils', as it was believed to cure seven ailments, including rickets, coughing, epilepsy, jaundice and shortness of breath.

# Sceartán
## *Tick*

If it has already engorged itself on the blood of a human or animal, it's known as BOLGADÁN. SNIODH is a nit and SNEÁCHÁN refers to someone with nits in their hair. The picking of vermin from clothes by hand is AISCEADH.

# Oiread na fríde
## *Tiny amount*

Refers to a tiny amount of a thing or almost nothing. FRÍD is a mite or a flesh-eating worm. AR MHAITH LEAT PÍOSA CÍSTE? ('Would you like a piece of cake?') SEA, OIREAD NA FRÍDE ('Yes, a mite's amount').

# Lus an tóiteáin
## *Houseleek*

A low-growing succulent plant with rosettes of tufted leaves. The name means 'plant of the destructive fire', since the plant was believed to protect against fire and lightning. Houseleeks were encouraged to grow in thatch roofs, and roofs of other materials had spaces left in them for the plant to grow. This tradition stretches back to the Roman Empire and possibly further still. The plant is also called TIN-EAGLACH ('fire-fear'), from the idea that fire was afraid of it.

# Creabhar
## *Gadfly*

A furry bumblebee-like insect that has no mouth. It doesn't need one, as the female lives only long enough to lay her eggs in the hairs near a cow's hooves. Later, a larva hatches and is basically one big mouth: it eats its way straight up through the cow towards the backbone, growing so fat that it creates lumps on the cow's back, called BHIARSÚIL. To avoid a gadfly trying to lay her eggs, a cow will perform crazed buck-leaping and even hurl herself over a cliff or into a lake. This behaviour is known as FÍBÍN.

# Uachtar go tóin
## *Butterwort*

A flesh-eating plant with broad green leaves covered in tiny glandular hairs that secrete sticky mucilage that acts like flypaper. It digests the flies that it catches. Its name translates as 'cream right to the bottom'. It's also called IM-GO-HUILLEANNA ('butter to the elbows'). These names don't refer to its greasy, buttery leaves that catch flies but to the belief that it was such a powerful plant that it could ward off witches or fairies who might spoil milk in the dairy by stealing the goodness from it.

It was so powerful that even if a cow ate butterwort in the field, the milk would be protected. Indeed, if a person ate cheese made from this milk, they too would be protected. It's most commonly called BODÁN MEASCÁIN, which can be translated as 'plant of the heap of moulded butter'.

# Snáthaid an phúca
## *Daddy-long-legs*

Translates directly as 'pooka's needle'.
SNÁTHAID AN DIABHAIL is also used,
which means 'the devil's needle' and can
also refer to a dragonfly. Another word
for dragonfly is SNÁTHAID MHÓR ('big
needle').

# Adharc an phúca
## *Stinkhorn mushroom*

The Irish name translates as 'the pooka's horn', as this nasty-smelling, alien-like mushroom appears suddenly from an egg-shaped form in the ground into a tall pillar with a white honeycomb top. When it emerges, the cap is covered with a sticky substance that is irresistible to flies. The mushroom definitely looks a bit demonic and otherworldly.

# Cáis an phúca
## *Giant puffball mushroom*

The birch polypore (a bracket-like mushroom that grows on the sides of birch trees). Because mushrooms appear mysteriously at night from deep within the ground, they were thought to be the creations of the pookas, evil spirits who crawl around on all fours. The name of the mushroom means either 'pooka's cheese' (CÁIS AN PHÚCA), because the spongy white mushrooms look and smell a bit like old cheese, or CAISE AN PHÚCA, which means 'pooka's stream (or current)'.

# Coileach
## *Rooster*

The word for its constant crowing is MIONGHLAO, and a rooster that crows regularly at the same hour every morning is a TRÁTHAÍ. COILICHÍN is a cockerel or an argumentative little person. The phrase TÁ COILICHÍN UIRTHI ('She has a little cockerel on her') means that she's riled and ready to fight. The Old Irish term MAGH NA HATHGABALA referred to the distance that a cockcrow could be heard from across a plain.

# Deargadh an dá néall
## *First light*

Literally 'the reddening of the two clouds'.

# Fás aon oíche
## *Mushrooms*

FÁS AON OÍCHE is a general name for mushrooms that translates as 'growth of one night', a reference to the amazing ability of mushrooms to appear fully formed in just a few hours. They are not really 'growing' so much as swelling up with water into the shape of a mushroom, as a sponge would. If they were growing through cell division, it would be a lot slower. Mushrooms are the reproductive organs (fruiting bodies) of a vast fungus network beneath the soil that takes years to develop. CUPÁN DRÚCHTA ('cup of early-morning dew') is another general name for mushrooms. BEACÁN is another, though you can also use MUISIRIÚN and BOLG BUACHAILL, which translates as 'cowherd's bubble (or belly or blister)' and BOCÁN, which also means 'goblin' and 'he-goat'.

# Breacadh an lae
## *Daybreak*

BREACADH means the brightening of the day. This time of day is followed by BÁNÚ AN LAE ('whitening of the day'). And as the sun rises further, it is FÁINNE GEAL AN LAE ('first bright ring of daylight'), then SÚIL AN LAE ('eye of the day') and, finally, ÉIRÍ NA GRÉINE, the sunrise itself.

# Maidneachan
## *Becoming morning or dawning*

You can also use BODHRÁNACHT AN LAE or LÁCHAN LAE, or you can use the verbs BÁNAIGH, FÁINNIGH, LÁIGH and FOINSIGH.

# HOW TO SAY
# THE IRISH WORDS
# IN THIS BOOK

The words collected in this book have been chosen for the insights they offer into our world and the ways that human consciousness engages with its surroundings. But as a book it lacks the ability to make the words audible, which is regrettable, as for many of them their sound is almost as evocative as their meaning. It's the resonance of the words in the air and reverberating through our bodies that makes such an impact.

So, I've attempted to create a rough pronunciation guide. I am neither a linguist nor phoneticist, so please forgive any inaccuracies. Furthermore, there is no single correct pronunciation for most Irish words, as there are at least three major dialects and many minor ones – each of which has different pronunciations for many words.

My name, for example, can be pronounced mon-a-KHAWN or MON-khan or mon-CON or MON-a-hun or man-a-HAN. All are correct. I was reared in Dublin, but my Irish is from the Munster dialect of West Kerry, hence I tend to emphasise the second syllable of words. But I've also toned down aspects of my Munster dialect to be more comprehensible to others. The following pronunciation guide is based on the way I say the words; others will pronounce them differently. Embrace the diversity. You can hear many of these words being pronounced in all three major dialects at www.teanglann.ie/en/fuaim.

## A

á róstadh . . . . . . . . . . . . aw ROW-stuh
adharc an phúca . . . . . . EYE-yerk uhn FOO-kah
ag ithe . . . . . . . . . . . egg I-huh
ag sclimpireacht . . . . . . egg SCLIMP-er-ukht
agh alla . . . . . . . . . . . EYE-OLL-eh
agus . . . . . . . . . . . . OG-us
aisceadh . . . . . . . . . . ASH-kuh
an fad . . . . . . . . . . . . uhn FOD
arcán . . . . . . . . . . . . ARE-kawn
asarlaíocht . . . . . . . . . . oss-er-lee-uckt
ascar . . . . . . . . . . . . OSC-er
athair thalún . . . . . . . . AH-er hol-OON

## B

bainirseach . . . . . . . . . BWIN-er-shuck
bainne . . . . . . . . . . . BON-nyu
bainne bó bleachtáin . . . BAH-nyeh bow BLAKH-taw-in
báire . . . . . . . . . . . . . BAHW-reh
bánaigh . . . . . . . . . . . BAWN-ig
bánú an lae . . . . . . . . . BAWN-OO uhn lay
bás dorcha . . . . . . . . . BAWSS durkheh
beacán . . . . . . . . . . . BACK-awn
beag . . . . . . . . . . . . BYUG
beainín uasal . . . . . . . . BANN-een OO-sel
bearach . . . . . . . . . . . BAHR-uhkh
beatha . . . . . . . . . . . BAH-hah
bhfuil . . . . . . . . . . . WILL
bhiarsúil . . . . . . . . . . VIR-sool
bior . . . . . . . . . . . . . BIRR
biorán suain . . . . . . . . BIRR-awn SUE-in
bior-rósanna . . . . . . . . BIRR-ROE-sunna
bleachtach . . . . . . . . . BLAHKH-tuhkh
bléineach . . . . . . . . . . BLAYNE-uhkh
bó shamhraidh . . . . . . . BOE HOW-rig
bó thórmaigh . . . . . . . . BOE HOAR-mig
bocán . . . . . . . . . . . . buck-AWN
bodachán . . . . . . . . . . BUD-uhkh-AWN
bodalán . . . . . . . . . . . BUD-ul-AWN
bodán meascáin . . . . . . . bud-AWN mask-AWN
bodhránacht an lae . . . . BAU-RAWN-ukht uhn LAY
bodóg . . . . . . . . . . . . bud-OGUE
bóín dé . . . . . . . . . . . BOW-een JAY
boinín . . . . . . . . . . . . bwin-EEN
bolg buachaill . . . . . . . . BULL-ug BOO-khill

bolgadán . . . . . . . . . . . BULL-geh-DAWN
borradh na sailí . . . . . . BURR-ah nah SAHL-ee
bradaí . . . . . . . . . . . . BRAD-ee
bradán . . . . . . . . . . . . brad-AWN
bradánach . . . . . . . . . . brad-AWN-uhkh
branaireacht . . . . . . . . . BRAN-er-uhkht
braobaire . . . . . . . . . . BRAY-burrah
breacadh an lae . . . . . . BRACK-uh uhn LAY
bró éisc . . . . . . . . . . . BRO AY-shk
brobh . . . . . . . . . . . . BROV
brobhaíl . . . . . . . . . . . brov-EEL
bruth . . . . . . . . . . . . bruh
buaircín . . . . . . . . . . . boor-KEEN
bual . . . . . . . . . . . . . BOO-el
bual-lile . . . . . . . . . . . BOO-el-LIL-eh
buarach . . . . . . . . . . . BOOR-ehkh
buicmín . . . . . . . . . . . BWICK-meen

## C

cá . . . . . . . . . . . . . . KAH
cadóg . . . . . . . . . . . . kad-OGUE
cailín bán . . . . . . . . . . kol-EEN BAWN
cailleach oíche . . . . . . . kwil-AHCH EE-heh
cáis an phúca . . . . . . . . KAWSH uhn FOO-kah
caise an phúca . . . . . . . KASH-ah uhn FOO-kah
cam an ime . . . . . . . . . KAUM uhn IM-eh
cam reilige . . . . . . . . . KAUM REL-igeh
capall . . . . . . . . . . . . KOP-ull
capánach . . . . . . . . . . kop-AWN-ahch
carbhán . . . . . . . . . . . kar-a-VAWN
carbhán carraige . . . . . kar-a-VAWN KARR-ig-eh
carria . . . . . . . . . . . . KARR-ee-ya
cat crainn . . . . . . . . . . KOT KREEN
ceann cait . . . . . . . . . . KY-OWN QUIT
ceannann . . . . . . . . . . KYAN-un
cearc Mhuire . . . . . . . . KYARK WIR-eh
ceart-aos . . . . . . . . . . KYART-ace
ciaróg na mbeannacht . . kir-OGUE nuh MAN-uhkht
cíoch an róin . . . . . . . . KEE-uhkh uhn ROW-in
císte . . . . . . . . . . . . . KEE-shteh
cladach rónta . . . . . . . KLAD-ahch ROWN-teh
clamhrán . . . . . . . . . . klou-RAWN

cleitire . . . . . . . . . . . . KLEH-tireh
clibistín . . . . . . . . . . . KLIB-ish-TEEN
clíthseach . . . . . . . . . . KLEE-shekh
cluich . . . . . . . . . . . . KLIH
cluiche gainéad . . . . . . KLIH-eh gan-AID
cnámharlach mairte . . . KNAWR-lukh MAR-teh
cnúdóg . . . . . . . . . . . KNOO-DOGUE
coigeal na mban sí . . . . KWIG-el nuh MON SHEE
coileach . . . . . . . . . . . KWILL-ahch
coilichín . . . . . . . . . . . KWILL-ikh-EEN
colann . . . . . . . . . . . . KULL-un
collphoc . . . . . . . . . . . KULL-fuc
comhla bhreac . . . . . . . KOWLA vrack
copsaí . . . . . . . . . . . . KUP-see
corránach . . . . . . . . . . kur-AWN-uhkh
costa . . . . . . . . . . . . KUS-teh
cráin . . . . . . . . . . . . . KRAW-in
cráin róin . . . . . . . . . . KRAW-in ROW-in
cráinbheach . . . . . . . . . KRAW-in-VAHCH
cráinseangán . . . . . . . . KRAW-in-shan-GAWN
crampánach . . . . . . . . kram-PAWN-ekh
crann creathach . . . . . . KROWN KRAH-hekh
crann-nasc . . . . . . . . . KROWN-nosk
creabhar . . . . . . . . . . KREU-wer
creathán . . . . . . . . . . krah-OWN
crobh préacháin . . . . . . KROV PRAY-ukh-AWN
croí . . . . . . . . . . . . . KREE
crúbóg . . . . . . . . . . . KROO-BOGUE
cú dobhráin . . . . . . . . KOO du-RAW-in
cuán mara . . . . . . . . . koo-AWN MAR-eh
cuasnóg . . . . . . . . . . . koos-NOGUE
cuasóg . . . . . . . . . . . koos-OGUE
cúbach . . . . . . . . . . . KOO-bukh
cudal méara . . . . . . . . KUD-al MAY-er-ah
cufróg . . . . . . . . . . . kuf-ROGUE
cupán drúchta . . . . . . . kup-AWN DREW-khteh

## D

dair . . . . . . . . . . . . . DAHR
daireach . . . . . . . . . . . DAHR-ekh
dallóg . . . . . . . . . . . . doll-OHWG
dallóg róin . . . . . . . . . doll-OHWG ROHW-in

dam conchaid . . . . . . . . . DAM CON-khid
damh . . . . . . . . . . . . . . DAV
damhán alla . . . . . . . . . . doo-WAWN OLL-eh
damhán uisce . . . . . . . . . dow-AWN ish-keh
damhra . . . . . . . . . . . . . DAV-rah
daróg . . . . . . . . . . . . . . dar-OWG
dathú . . . . . . . . . . . . . . dah-HOO
deannach . . . . . . . . . . . . DYAN-ehch
dearg . . . . . . . . . . . . . . DYAR-UG
deargadh an dá néall . . . DYAR-geh uhn DAW NAIL
deilf . . . . . . . . . . . . . . . DELF
diúilín . . . . . . . . . . . . . dew-LEEN
dobharchú . . . . . . . . . . . DOUR-khoo
dobhar-each . . . . . . . . . . DOUR-AKH
dobhrán . . . . . . . . . . . . doo-RAWN
dreoilín ceannbhuí . . . . DROH-LEEN kyo-wn-VEE
dreoilín easpaig . . . . . . DROH-LEEN ASS-pig
dreoilín teaspaigh . . . . . DROH-LEEN TASS-pig
dreollán . . . . . . . . . . . . DROH-LAWN
dril . . . . . . . . . . . . . . . DRIL
driseog . . . . . . . . . . . . . drish-OWG
droimeann . . . . . . . . . . . DRIM-un
drúchtín . . . . . . . . . . . . DREW-uhk-TEEN
drúi donn . . . . . . . . . . . DREE down
duilleoga báite . . . . . . . dill-OWG-eh BAW-teh

# E

each . . . . . . . . . . . . . . . AKH
eachra . . . . . . . . . . . . . AKH-ruh
éan róin . . . . . . . . . . . . AIN ROHW-in
earc luachra . . . . . . . . . ARK LOO-khrah
earc sléibhe . . . . . . . . . ARK SHLAY-veh
eascann . . . . . . . . . . . . . ASK-en
easóg . . . . . . . . . . . . . . ass-OWG
éigne . . . . . . . . . . . . . . AY-gneh
éigneach . . . . . . . . . . . . AY-gnekh
eireaball cait . . . . . . . . ER-ah-boll QUIT
éirí na gréine . . . . . . . . EYE-REE nuh GRAY-neh
eo . . . . . . . . . . . . . . . . OH

# F

fáinne geal an lae . . . . . . FAWN-nye gal uhn LAY
fáinnigh . . . . . . . . . . . . FAWN-ig
falaire . . . . . . . . . . . . . FAL-era
falaireacht . . . . . . . . . . FOLL-er-ukht
falscaithe . . . . . . . . . . . FAL-ske-huh
faoileán . . . . . . . . . . . . FWEEL-AWN
faoileánach . . . . . . . . . . FWEEL-AWN-ekh
faoileanda . . . . . . . . . . . FWEEL-en-deh
faol . . . . . . . . . . . . . . . FWAOL
faolchú . . . . . . . . . . . . . FWAOL-COO
fás aon oíche . . . . . . . . FAWS ay-un EE-heh
feadán . . . . . . . . . . . . . fad-AWN
feam . . . . . . . . . . . . . . FAM
feannóg . . . . . . . . . . . . fan-OWG
fearb . . . . . . . . . . . . . . FAHR-ub
fearbán . . . . . . . . . . . . fahr-BAWN
fearn . . . . . . . . . . . . . . FAHRN
fearnóg . . . . . . . . . . . . fahrn-OWG
fearphoc . . . . . . . . . . . . fahr-FOC
feimide . . . . . . . . . . . . FEM-idj-eh
feimín . . . . . . . . . . . . . fe-MEEN
feimíneach . . . . . . . . . . fe-MEEN-ekh
féirín sí . . . . . . . . . . . . fair-EEN SHEE
fia . . . . . . . . . . . . . . . . FEE-eh
fia rua . . . . . . . . . . . . . FEE-ah ROO-ah
fiamhíol . . . . . . . . . . . . FEE-eh-VEAL
fíbín . . . . . . . . . . . . . . FEE-BEEN
fíneach . . . . . . . . . . . . . FEE-nekh
flannóg . . . . . . . . . . . . flonn-OWG
fleá . . . . . . . . . . . . . . . FLAW
foinsigh . . . . . . . . . . . . FWIN-shig
foirgneamh . . . . . . . . . . FUYR-gnuv
fraecsáil . . . . . . . . . . . . fraykh-SAW-il
fraoch . . . . . . . . . . . . . FRAYKH
fraochdhaite . . . . . . . . . FRAYKH-ghait-eh
fraoigh . . . . . . . . . . . . FRAY-ig
fríd . . . . . . . . . . . . . . . FREEDJ
fuairneach . . . . . . . . . . . FOOR-nekh
fuarghaoth . . . . . . . . . . FOOR-gway
fuinseog . . . . . . . . . . . . fween-SHOWG
fuipín . . . . . . . . . . . . . fwip-EEN

# G

gabhairín . . . . . . . . . . . GOW-er-EEN
gabhar . . . . . . . . . . . . . GAW-er
gabhar deorach . . . . . . . GAW-er DYOR-ekh

Gaeilge . . . . . . . . . . . . GWALE-geh
gan . . . . . . . . . . . . . . . GONE
gealóg bhuachair . . . . . . gal-OWG VOOK-er
gearrán . . . . . . . . . . . . gar-AWN
gillín róin . . . . . . . . . . gill-EEN ROW-in
gionán . . . . . . . . . . . . . gyun-AWN
glas . . . . . . . . . . . . . . . GLOSS
glasdair . . . . . . . . . . . . GLOSS-DAHR
glasóg . . . . . . . . . . . . . gloss-OWG
go léir . . . . . . . . . . . . . guh LAYER
gob uirthi . . . . . . . . . . . GUB UHR-hee
gormánach . . . . . . . . . . gur-MAWN-ehkh
grafainn . . . . . . . . . . . . GRAF-iyn
graí . . . . . . . . . . . . . . . GREE
graifne . . . . . . . . . . . . GRAF-nyeh
graifneach . . . . . . . . . . GRAF-nekh
gráinneog . . . . . . . . . . . grawn-OWG
gráinneog thrá . . . . . . . . grawn-OWG HRAW
gráinneogach . . . . . . . . . grawn-OWG-ekh
greamanna . . . . . . . . . . GRAM-un-eh

# H

hob amach . . . . . . . . . . HUB um-AKH
hobaireacht . . . . . . . . . . HUB-er-ekht
huít, huít . . . . . . . . . . . WHO-EET, WHO-EET
hurais . . . . . . . . . . . . . HUR-ish

# I

ialtóg . . . . . . . . . . . . . . eel-TOWG
ime . . . . . . . . . . . . . . . IM-eh
im-go-huilleanna . . . . . . IM guh HILL-uhn-eh
íochtairín . . . . . . . . . . . EE-ekh-ter-EEN
iomas gréine . . . . . . . . . UM-es GRAY-neh
iomghaothach . . . . . . . . um-GWEE-hekh
iora glas . . . . . . . . . . . . IYR-eh GLOSS
iora rua . . . . . . . . . . . . IYR-eh ROO-EH

# L

labhairt . . . . . . . . . . . . LAUW-ert
lacha . . . . . . . . . . . . . . LOUGH-eh
lacha mhásach . . . . . . . LOUGH-eh VAW-suhkh
láchan lae . . . . . . . . . . LAWKH-un LAY
lachar . . . . . . . . . . . . . LOKH-er
lachnach . . . . . . . . . . . . LOKH-nekh

| | | | |
|---|---|---|---|
| láigh | LAH-ig | madra rua | MOD-reh ROO-a |
| langán | lan-GAWN | madra uisce | MOD-reh ISH-ka |
| lasair choille | LOSS-er khill-eh | maidneachan | MY-nukh-un |
| leathadh | LAH-heh | maighre | MYE-reh |
| leathóg | LAH-HOWG | maighreach | MYE-rekh |
| leathóg bhallach | LAH-HOWG VALL-ekh | maitheas | MAH-hus |
| leathóg bhán | LAH-HOWG VAWN | maoilín | MWAY-leen |
| leathóg mhuire | LAH-HOWG WIRR-eh | mara | MOR-eh |
| | | marc | MARK |
| leis | LESH | marcach ar sheangán | mar-KAHCH er hang-AWN |
| liath | LEE-eh | másach | MAW-sukh |
| locha | LUKH-eh | máthair shúigh | MAW-hir WHO-ig |
| loilíoch | lull-EE-ukh | meacan dubh | MACK-en DUV |
| luaineach | LOO-in-ekh | meannán aeir | man-AWN AIR |
| luibh na ndaitheacha | LIV nuh NAH-huk-ah | méaracán an diabhail | MAY-er-a-KAWN uhn DEE-yal |
| luibh na seacht ngábh | LIV nuh SHOCKHT NGAWV | méaracán daoine marbh | MAY-er-a-KAWN DEE-neh MAR-iv |
| lus an bhainne | LUSS uhn VON-yeh | meig | MEG |
| lus an tóiteáin | LUSS uhn TOWTCH-AWN | meigeall | MEG-el |
| | | meigeallach | MEG-el-ekh |
| lus gan athair gan mháthair | LUSS gon A-her gon VAW-her | méiríní madra rua | MAYR-een-ee MOD-reh ROO-a |
| lus mór | LUSS MORE | méiríní na mban sí | MAYR-een-ee nuh MON SHEE |
| lus na gcnámh briste | LUSS nuh GNAWV BRISH-teh | minseach | MIN-shekh |
| lus na mban sí | LUSS nuh MON SHEE | míog | MEE-owg |
| lus na saighdiúirí | LUSS nuh side-OOR-ee | míol | MEE-ul |
| | | míol mór | MEE-ul MORE |
| lus na seacht mbua | LUSS nuh SHOCK-HT MOO-eh | míoltóg | MEE-ul-TOEWG |
| | | mionghlao | MYUN-ghlay |
| | | móin | MOE-in |
| | | monghar | MUN-ghur |
| | | monghar na laoch | MUN-ghur nuh LAY-ekh |

## M

| | |
|---|---|
| mac aille | MOCK AL-yeh |
| mac an daba | MOCK uhn DOB-eh |
| mac fhalla | MOCK OL-a |
| mac imrisc | MOCK IM-rishk |
| mac samhla | MOCK SOW-la |
| mac tíre | MOCK TEE-reh |
| macalla | MOCK-OLL-ah |
| macasamhail | MOCK-ah-SOW-il |
| mac-chailleach | MOCK-khyle-ekh |
| macleabhar | MOCK-LAUW-er |
| macsaithe | MOCK-HIGH-eh |
| madra allta | MOD-reh OWL-teh |
| madra crainn | MOD-reh KREEN |
| madra dearg | MOD-reh DYAR-ig |

| | |
|---|---|
| muc | MUCK |
| muc mhara | MUCK WOR-ah |
| mucachán | muck-a-KAWN |
| muirbheach | mwir-VAKH |
| múirling | MOOR-ling |
| muisiriún | mush-ROON |

## N

| | |
|---|---|
| na circe | nuh KIR-keh |
| na deirfiúiríní | nu driff-OOR-EEN-ee |
| na gcon | nuh GUN |
| na smóilíní | nuh smole-EEN-EE |
| naosc | NAY-esk |

| | |
|---|---|
| naoscach | NAY-esk-ekh |
| nead | NYAD |
| neadaireacht | NYAD-er-ekht |
| neanta | NYAN-teh |
| ní | NEE |
| níl | NEEL |

## O

| | |
|---|---|
| ó bearradh na rónta | OH BAHR-ekh nuh ROEN-te |
| oiread na fríde | IR-id nuh FREE-deh |
| oisín | ush-EEN |
| oisín róin | ush-EEN ROW-in |
| os | US |
| oscartha | US-ker-ha |

## P

| | |
|---|---|
| páirc | PAW-irk |
| páirc éisc | PAW-irk AYSHK |
| peall | PAL |
| péist capaill | PAY-sht COP-el |
| pilibín | PILL-a-BEEN |
| pilibín míog | PILL-a-BEEN MEE-owg |
| píosa | PEE-sah |
| plúirín | PLOOR-een |
| poc | PUCK |
| pocán | puck-AWN |
| poibleog | pwib-LOWG |
| póicíní | POE-KEEN-ee |
| póicíní locha | POE-KEEN-ee LUKH-ah |
| póiríní seangán | POUR-EEN-ee shan-GAWN |

## R

| | |
|---|---|
| raideog | radj-OWG |
| rail | RAL |
| rannaireacht | RONN-er-ekht |
| rannán | ronn-AWN |
| raspa | RAS-pah |
| ratamas | RAT-em-us |
| ráth | RAW |
| ráth scadán | RAW skod-AWN |
| ráthaigh | RAW-hig |
| rí rua | REE ROO-a |

| riabhach | REE-vokh |
|---|---|
| riathróir | REE-ah-RORE |
| ró-fhos | ROW-uss |
| rois | RUSH |
| róngáe | ROWN-gah |
| ropadh | RUP-eh |
| ros lachan | ROS LOKH-en |
| rósanna | ROWS-uhn-eh |
| roth buaile | RUH BUWIL-ye |
| rua | ROO-ah |
| rua-ghaoth | ROO-ah-GWAY |
| ruaim | ROO-im |
| rúplach | ROOP-lekh |

## S

| saighneán | SYE-NAWN |
|---|---|
| sail éille | SAL AY-leh |
| sailchuach | SAL-KOOW-ek |
| saileog | SAL-owg |
| sailleach | SAL-ekh |
| saol | SAY-el |
| saothar | SAY-her |
| scadán | skod-AWN |
| scadán caoch | skod-AWN QAY-ekh |
| scadán gainimh | skod-AWN GANN-iv |
| scadán láibe | skod-AWN LAW-ib-eh |
| scadán na bpis | skod-AWN nuh BISH |
| scamal | SKOM-el |
| scaraoid | skair-AID |
| sceartán | shkirt-AWN |
| sciathán leathair | shki-HAWN LAH-ir |
| scimeáil | shkim-ALL |
| sclimpireacht | SHKLIMP-er-ekht |
| scodalach | SKUD-el-ukh |
| scráib | SKRAW-ib |
| scréachóg reilige | SHCRAY-KHOWG REIL-ig-eh |
| seabhac | SHAU-ek |
| seabhac gaoithe | SHAU-ek GWEE-he |
| seabhaic | SHAO-ik |
| seabhcán | shau-KAWN |
| seabhcóir | shau-KORE |
| seabhcóireacht | SHAU-KORE-ekht |
| seabhcúil | SHAO-KOO-yl |
| seacht n-óg na coille | SHOCKHT NOGUE nuh KWILL-eh |
| seacht n-óg na farraige | SHOCKHT NOGUE nuh FORR-ig-eh |

| séamas rua | SHAY-mus ROO-ah |
|---|---|
| seangán | shan-GAWN |
| seile cuaiche | SHELL-a COO-ih-eh |
| seilín cuaiche | SHELL-EEN COO-ih-eh |
| seitreach | SHET-rekh |
| siansán | sheen-SAHWN |
| sionnach | SHUN-ekh |
| sirtheoireacht | SHIR-HORE-ekht |
| slánlus | SLAWN-luss |
| slánlus mór | SLAWN-luss MORE |
| sliabh | SHLEE-ev |
| sliomach | SHLUM-ekh |
| smaois | SMWEE-sh |
| sméar | SMAY-er |
| smuga | SMUG-ah |
| smugachán | SMUG-a-khawn |
| smugairle róin | SMUG-er-leh ROW-in |
| smúrthacht | SMOOR-hakht |
| snag | SNOG |
| snáthaid an diabhail | SNAW-hud uhn DEE-yl |
| snáthaid an phúca | SNAW-hud uhn FOO-kah |
| snáthaid mhór | SNAH-hud VOOR |
| sneáchán | SHNAW-a-khawn |
| sniodh | SHNEE |
| soirbheas | SIYR-vus |
| sop píce | SUP PEEKA |
| sopaireacht | SUP-er-ukht |
| sopóg | sup-OWG |
| spágaire | SPAWG-era |
| spangaire | SPONG-ir-eh |
| spáráil | SPAW-RAW-yl |
| speathán | spa-HAWN |
| speathánach | spa-HAWN-ekh |
| speig | SPEG |
| speig neanta | SPEG NYAN-teh |
| speirsín | sper-SHEEN |
| spideog | SPID-owg |
| sprochaille | SPRUCK-il-la |
| sraith éadaí | SRAH AIDH-ee |
| stadhan | STYNE |
| staga | STOG-ah |
| staigín | staigg-EEN |
| stuaicín | stook-EEN |
| súil an lae | SOOYL uhn LAY |

## T

| talmhóg | tal-VOGUE |
|---|---|
| tarbh | TAR-iv |
| tarbhshaithe | TAR-iv-HEIGH-he |
| teadhall | TAH-yll |
| teanga | TAHN-geh |
| tháinig | HAWN-ig |
| thug | HUG |
| tin-eaglach | TIN-OG-lekh |
| tionlacan | TYUN-lekh-un |
| toch toch | TUCKH TUCKH |
| tóithíní | TOE-HEEN-ee |
| tóithíní muca mara | TOE-HEEN-ee MUCK-a MAR-a |
| tonn | TOWN |
| tonóg | tun-OWG |
| tormán | tur-MAWN |
| tráthaí | TRAW-HEE |
| tuaithleasóg | TOO-yl-ass-OWG |

## U

| uachtar go tóin | OOK-ter guh TOW-in |
|---|---|
| uaine | OO-nya |
| ulchabhán | ULL-khuv-AWN |
| urchall | UHR-khel |

# Acknowledgements

I'd like to thank all the people who made this book possible: Manchán Magan for his words; Graham Thew for his wonderful design; Deirdre, Aoibheann, Teresa, Bartek and everyone at Gill Books for being so supportive; my wife Dara Ní Bheacháin for her keen designer's eye and always excellent advice; my daughters Róisin and Ailsa, the loveliest, rarest creatures in Ireland; and my parents, Peggy and Willie Doogan. Thank you all so much. *Steve*

Gill Books
Hume Avenue
Park West
Dublin 12
www.gillbooks.ie

Gill Books is an imprint of M.H. Gill & Co.

Text © Manchán Magan 2021
Illustrations © Steve Doogan 2021

9780717192557
Copy-edited by Ruairí Ó Brógáin
Proofread by Fidelma Ní Ghallchobhair
Designed by Graham Thew
Printed by Printer Trento, Italy

This book is typeset in 15 on 18pt Minion Pro

5 4 3 2 1